Apple Tv User Guide

A concise Practical Guide with Tips and Tricks to Maximizing the New tvOS 14 with illustrative screen shots

By

Philips Keith

Table of Contents

Part 1

Part 2

Part 3

Step By Step Instructions To Adjust TV
App Video Download Settings On Iphone 21

Part 4

Part 1

Tvos 14

tvOS 14 is the latest version of the software to run on Apple TV. Available now.

Every year, Apple releases an update for Apple TV. More often than not, tvOS isn't totally modified, yet Apple consistently rolls out certain improvements that may appear to be little on a superficial level, however are in reality large upgrades. So what's happening for the Apple TV OS? Peruse on to discover!

So what are the new features of the Apple TV update?

The greatest changes in tvOS 14 include a few updates in the engine to improve performance, however we know you're here for the great stuff. Here are the new highlights in the current year's Apple TV update.

Multi-User Support For Apple Arcade

Prior to tvOS 14, Apple TV did not support multi-user support for syncing with Apple Arcade. You would now be able to have various client profiles, each associated with Apple Arcade on a similar Apple TV, and your promotion will be saved and prepared for when you play once more. At the point when you switch profiles, you'll see your game improvement, friends' rundown, and more.

VPN Deals: $ 16 lifetime permit, $ 1 month to month plans, and more.

Support For Xbox Elite 2 And Xbox Adaptive Controller

Since 2019, Apple TV supports Xbox One and PlayStation 4 regulators and now with tvOS 14 Xbox Elite 2 is likewise supported, and above all, the Xbox Adaptive Controller can be utilized on Apple TV. This is a stunning expansion to games on Apple TV for accessibility,

Screen Saver Themes

In tvOS 14, Apple has added the capacity to choose plane screensaver themes to show when your Apple TV is inactive. Through another alternative accessible in the Settings menu, you would now be able to look over the following themes: Landscape, Earth, Underwater, Cityscape. This setting is win big or bust, so you can't choose singular screensavers within a theme.

Share 4K Videos In The Photos App

When review vacation photographs and recordings on your Apple TV, you can see those recordings in 4K from your iPhone or iPad through AirPlay with 4K HDR. Furthermore, if your family needs to send recordings to their Apple TV from their gadget, they'll likewise be shown in 4K.

Pip On All Content

In tvOS 14, you can exploit Picture-in-Picture in both the Apple TV application and some other application. So you can watch the news while you work out or keep streaming your TV shows as you look for another home on Zillow. PiP likewise works with AirPlay.

New Integration Between Homekit And Home View

In tvOS 14, we have full help for review pictures from HomeKit viable cameras. HomeKit viable cameras are incorporated with Apple TV and you can see them in the new Home view in Control Center. The underlying perspective in Control Center is a stripped down variant of the launcher application (so no. there is no launcher application on Apple TV yet) that permits you to get to your number one cameras and scenes that you have labeled in the home application on your iPhone or iPad

You can likewise get live doorbell warnings on your Apple TV for HomeKit Secure Video-empowered savvy cameras that incorporate the name of the individual at the entryway because of facial recognition.

You can ask Siri on Apple TV to show you the view from one of the cameras whenever and a brilliant PiP window will show up without intruding on your films or shows. Additionally, you can see the camera in full screen, and interestingly, HomeKit cameras on the big screen likewise incorporate sound with tvOS 14.

Airpods Audio Sharing On Apple TV

At the point when you and your partner settle down for the evening yet don't have any desire to wake the children, you would now be able to set up two sets of AirPods or AirPods Pro as sound gadgets on tvOS 14.

You can begin that activity film without waking the child.

YouTube in 4K

I know, why has this not, at this point been an alternative? However, it at long last comes to tvOS 14. You will actually want to watch recordings in full 4K goal directly on your big screen on your 4K (or higher) TV. It's nearly time.

Privacy and Tracking

tvOS 14 also presents another following choice accessible in the protection settings. This setting permits clients to incapacitate outsider application movement following, true to form for iOS 14 preceding the component was postponed until one year from now. For what reason is it called tvOS14?

Apple changed to the tvOS SDK in 2015. It was as of late established on the iOS SDK. To stay as per the principal type of the iOS numbering. So notwithstanding the way that tvOS was in its first cycle in 2015, Apple called it tvOS 9. After five years, we are at the fourteenth arrival of an Apple TV working framework that was based on iOS, which has likewise seen its fourteenth release.

Is The Apple TV Update Available?

IS! If your Apple TV hasn't automatically updated yet, you can manually remove it when it's ready. Please note that the update may take a while, so plan your late-night activities to watch TV.

How to update Apple TV

What amount will it cost?

Literally nothing. Apple has quit charging for programming updates of OS X Mavericks. We never pay for iPhone, iPad, Apple Watch or Apple TV refreshes.

Will My Apple TV Work With Tvos 14?

Practically sure. However long your Apple TV is running tvOS, you can run tvOS 14 on it. It's any Apple TV before 2015.

Apple TV gadgets from 2014 and prior adaptations run Apple TV programming that utilizes a channel-style content establishment framework rather than applications. In case you don't know which Apple TV you have, discover an App Store. On the off chance that you don't have an App Store, you have a more established Apple TV that doesn't uphold tvOS.

How Can I Get Tvos 14 On My Apple TV?

Your Apple TV ought to request that you have an update accessible to introduce, or it might refresh manually for the time being..

Open the app settings on your Apple TV.

Click on System.

Snap on System. Snap Software Update.

From here, you can now begin the installation of Apple TV update.

At first sight

tvOS 14 is the latest version of tvOS with Picture in Picture, 4K YouTube video, multi-user gaming support, AirPods audio sharing, and more.

Part 2

Characteristics

Watch YouTube recordings in 4K

Backing for games saved by numerous clients

New help for Xbox regulators

Starter application for HomeKit

Share souind with AirPods

tvOS is the working system that spikes in demand for the fourth and fifth era Apple TVs, giving a simple to-explore TV seeing experience on the Apple link box.

With an exhaustive application store, tvOS supports downloading a wide range of applications and games that can be utilized on Apple TV, and the interface configuration puts content up front. Access what you need to see with Siri orders, Apple Remote, or the Remote application on iPhone and Apple Watch.

tvOS includes a few inherent applications like Photos to get to your photograph library, Apple Music, Podcasts, and Apple TV, an application that totals TV and film content from an assortment of sources, including the Apple TV + web-based feature.

The channels are included in the Apple TV app, so you can buy in and watch paid assistance content without hosting to open a third-get-together application.

Apple regularly adds new features to tvOS, and the tvOS 2020 update is tvOS 14. tvOS updates never make however many changes as iOS or macOS refreshes, yet there are still some critical new highlights significant.

tvOS 14 highlights support for various game regulators, so you can combine the Xbox Series 2 Elite Wireless Controller and Xbox Adaptive Controllers with your Apple TV

Multi-users support for games has been added, so each tvOS client can monitor their game levels, leaderboards, and welcomes while playing Apple TV games. Another alternative permits you to physically choose a screensaver family, so you don't need to go through irregular trading.

There is currently a Home section in the Control Center on Apple TV, which is an incredible method to control your HomeKit associated items directly on your TV. There is additionally an alternative to see the HomeKit camera pictures in

Big Screen.

Picture-in-Picture mode allows you to watch a film while you work out, watch a sporting event, or open a news application, and sound sharing help allows you to interface two arrangements of AirPods to an Apple TV so two individuals can tune in without upsetting. . AirPlay permits you to share 4K recordings from the Photos application and view them in full goal or view them in Picture-in-Picture mode and, interestingly, tvOS 14 allows you to see YouTube recordings in 4K, albeit this component is as yet being dispatched from YouTube

Apple delivered the update to tvOS 14 in September 2020. It very well may be introduced on fourth and fifth era Apple TV models.

Current Version

The current form of tvOS will be tvOS 14.4, delivered to the general population on January 26th. As indicated by Apple's delivery notes for the update, tvOS 14. 4 added general execution and security updates.

tvOS 14.4 follows tvOS 14.3, an update that additional help for Apple Fitness +, Apple's new membership based exercise experience that utilizes the Apple Watch to follow exercises, and another Fitness application shows up on Apple TV when completed, refreshed to tvOS 14.3

Apple has also delivered two beta forms of a forthcoming tvOS 14.5 update for developers and public beta testers for testing purposes. The update adds support for the new Xbox Series X and PlayStation 5 Dual Sense controllers.

New Features Added In Tvos 14

tvOS doesn't get as numerous updates and highlight changes as a portion of Apple's other working frameworks, yet Apple adds new highlights each fall in each refreshed rendition of tvOS.

tvOS 14, which is in beta testing at the present time and will be delivered later in 2020, highlights a small bunch of valuable .changes significant

.Picture in Picture

Picture-in-Picture mode allows you to watch films, TV shows, or applications in a little window toward the edge of the screen, while likewise accomplishing something different on Apple TV.

You can watch a film while you work out, watch a games match while you play, or open a news application to see the day's features without missing your number one TV show.

New Airpods Features

Apple TV sound sharing help allows you to connect two arrangements of AirPods to a solitary Apple TV so two individuals can tune in to a TV show or film without upsetting each other in the room.

Home Checks

Apple has added home controls to tvOS 14, permitting you to view and control some HomeKit-associated gadgets directly from the Apple TV Control Center. There is a choice to see HomeKit camera pictures on the bigger TV screen and they can likewise be seen in Picture-in-Picture mode.

4K Video Streaming

tvOS 14 allows you to watch YouTube recordings in 4K interestingly, adding support for YouTube's 4K video codec, yet YouTube still can't seem to refresh its application for the element to work. It also allows you to share 4K videos from the Full

Resolution Photos application or view them in Picture-in-Picture mode.

Multi-User Support For Games

On the off chance that you have different individuals who use Apple TV at home and love to play Apple Arcade, each tvOS client would now be able to monitor their individual game levels, leaderboards, and invites.

Game advancement is additionally saved per client with the goal that different individuals can play a similar game.

More Game Controller Support

tvOS 14 adds support for extra game regulators that can be matched with Apple TV to play tvOS games. Apple TV is viable with the Xbox Elite Wireless Controller Series 2 and the Xbox Adaptive Controller.

Screen Saver Options

There is a new option to choose a screen saver in tvOS 14. In previous versions of tvOS, screen savers could not be chosen manually and were displayed randomly.

tvOS 14 lets you choose a specific group of screensavers, such as ocean, space, or city, and Apple TV toggles between these screensavers.

Main Features Of Tvos

TV app

The Apple TV application is Apple's all inclusive resource for staying aware of the TV shows you love, discovering new substance, and getting proposals. The general application interface has segments for films, TV shows, sports, and children's substance at the top on both Apple TV and iPhone, while the library has a rundown of content you've purchased from iTunes.

"Watch now" with its "Next" include still shows up in the TV application, yet there is another AI based proposal motor that showcases customized content ideas for you dependent on what you need.

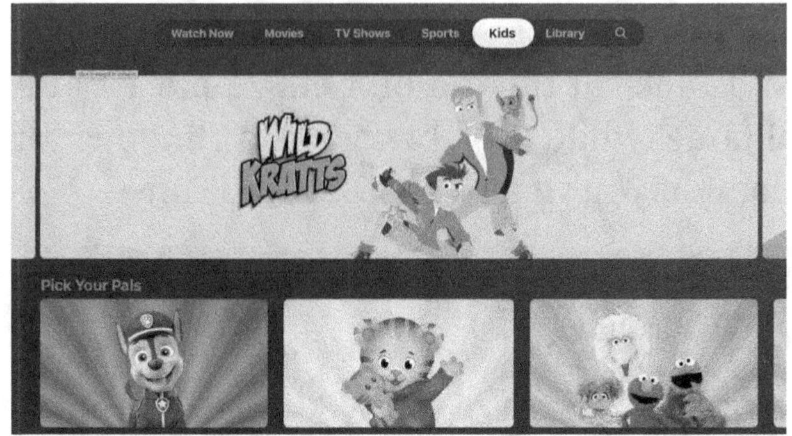

Up Next continues following what you're noticing so you generally recall which scene of a TV show you're watching or where you quit watching a film, while the "For You" suggestion highlight draws content from more than 150 streams. Applications including Hulu, Amazon Prime, DirecTV Now, PlayStation Vue, and anything is possible from that point. Notwithstanding the "For you" segment, the TV application likewise offers "For what reason did you watch ..." suggestions very much like Netflix.

Instructions to change the streaming quality settings of the TV application on iPhone

Part 3

Step By Step Instructions To Adjust TV App Video Download Settings On Iphone

Channels

The TV application incorporates a "Channels" area, which is a key help include presented by Apple in 2019. Channels are membership benefits that you can buy in to and watch inside the TV application without opening another application. So for instance, in the event that you discover a show you need to watch on your iPhone or Apple TV on Showtime, you can tap to buy in to Showtime directlyin the TV application,, at that point you can watch that show without leaving the application.

A portion of the supported channels incorporate CBS All Access, Starz, Showtime, HBO, Nickelodeon, Mubi, The History Channel Vault, Comedy Central Now, and AMC +. Apple has likewise begun offering uncommon packaged estimating to Apple TV + endorsers, and Apple TV + clients can pay $ 9.99 each month to watch CBS All Access and Showtime.

You will keep on accepting substance suggestions from administrations that are not part of the channels, so regardless of whether Hulu isn't something you can buy in to and watch in the TV application itself (you need to watch the Hulu content in the Hulu application). , you can in any case see the Hulu content ideas very much like the first TV application.

Accessibility

The TV application is accessible on iPhone, iPad, Apple TV, and Mac. Apple has likewise brought the Apple TV application to Roku and Amazon Fire TV, just as savvy TV contributions from organizations like Sony and Samsung.

Apple TV +

Apple TV + offers the entirety of Apple's unique TV shows and films, for example, "For All Mankind", "Dickinson", "Worker" and " The Morning Show ".

Apple TV + costs $ 4.99 each month and up to six relatives can share admittance to a membership utilizing Family Sharing. Mac is offering all clients who buy an iPad, iPhone, Mac, or Apple TV beginning September 10, 2019, a free one-year membership to Apple TV +.

To learn more about Apple TV +, be sure to check out our guide on Apple TV +.

Apple has in excess of two dozen unique TV shows and motion pictures underway, many highlighting prominent entertainers, entertainers, chiefs and makers. We have a total guide posting all the TV and film projects Apple has being developed accessible here.

Games

tvOS supports games designed for Apple TV, including those freely created for tvOS and accessible through Apple Arcade, Apple's $ 4.99/month membership administration for games. Macintosh Arcade gives admittance to over 100 games on iPhone, iPad, Apple TV and Mac, all without extra in-application buys or advertisement.

Games can be played with the Apple Remote, a game regulator designed for iOS, or a reassure game regulator, for example, the famous PlayStation DualShock regulator.

Siri

Siri on Apple TV is initiated by squeezing the devoted Siri button on the Siri far off and afterward talking an order. Siri can react to a wide assortment of solicitations on Apple TV, doing everything from offering film suggestions to uncovering entertainers on a TV show.

Like iOS, Siri can open applications and games and react to orders that are something other than content hunts. For instance, Siri can see sports scores, film times, climate, and stock status. Siri can likewise be utilized to change explicit settings with orders, for example, "Turn on cutting edge discourse", an element that expands exchange and mellow music and audio cues, or "Turn on subtitles" for subtitles.

Siri understands points and can react to look through dependent on themes like "Show me motion pictures from the 80s" or "Show me films with dinosaurs" or "Discover design ". documentaries". Siri can likewise tell when different points are remembered for a solitary order, for example, "Show me 1960s spy motion pictures" or "Show me 90s secondary school comedies."

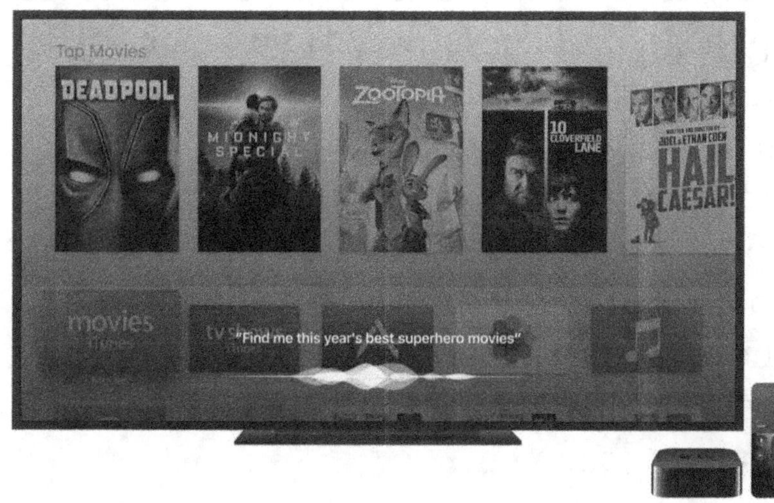

A Siri highlight called Live Tuning permits Siri to open live TV content in the application with orders like "Watch ESPN" or "Watch CBS" and Siri can likewise discover content inside the applications. For example, "discover me YouTube accounts with felines" dispatches the YouTube application and introductions relevant rundown things, similarly as a request, for instance, "find me comedies on Netflix"

Siri likewise works related to a framework wide pursuit include in tvOS that permits searches to show content from numerous applications like Netflix, iTunes, Hulu, HBO Go, Showtime, and the sky is the limit from there. Along these lines, on the off chance that you look for something like "Harry Potter", all the different applications where you can watch a Harry Potter film will be shown. A list of applications that support system-wide search is available in an Apple support document.

One of the best benefits of Siri is a feature that plays content. During a TV show or film, ask Siri "What did she just say?" or a comparable order and Siri returns 15 seconds and briefly turns on captions. Rewind and quick forward can likewise be performed with voice orders, for example, "Quick forward five minutes" or "Play from start".

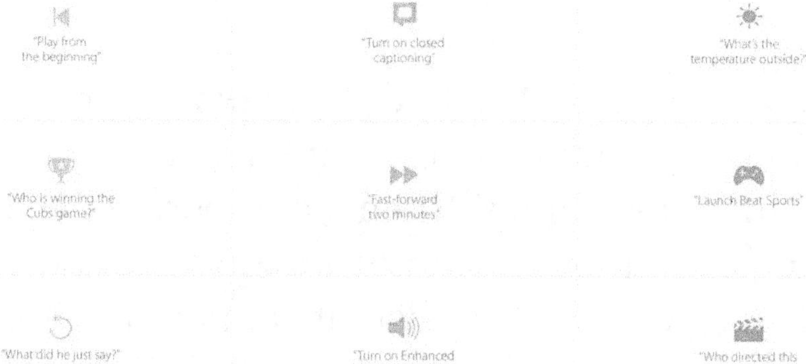

Siri shows cast information for a movie or TV show when prompted. Posing inquiries like "Who coordinated this film?" or "Who stars in this film?" shows a cast list. Siri can channel by cast, chief, date or age.

With controls that show extra substance, for example, in the background data, sports scores, or climate, the data shows up at the lower part of the Apple TV interface so you don't intrude on your on-screen show or film.. Tapping the remote opens the bottom bar in full screen, pausing the TV show or movie being played, and it's easy to switch between activities with a simple tap of the remote.

Mac TV, as iPad and HomePod, can go about as a home center point that associates with your HomeKit items and permits you to get to them distantly when you're away from home. For distant admittance to HomeKit gadgets, an Apple TV, iPad, or HomePod is required. Something else, HomeKit items possibly work when associated with a similar Wi-Fi network.

How To Use Password Autofill On Apple TV

The Most Effective Method To Utilize Secret Word Autofill On Apple TV And Tvos 12

The Most Effective Method To Utilize Secret Word Autofill On Apple TV And Tvos 12

Getting to applications and administrations on Apple TV can be baffling when exploring characters on the onscreen keyboard or shouting letters and numbers on the Siri Remote. To make it simpler to enter login certifications, Apple presented the Continuity Keyboard, an iOS highlight that permits clients to enter text on their Apple TV utilizing their iPhone or iPad.

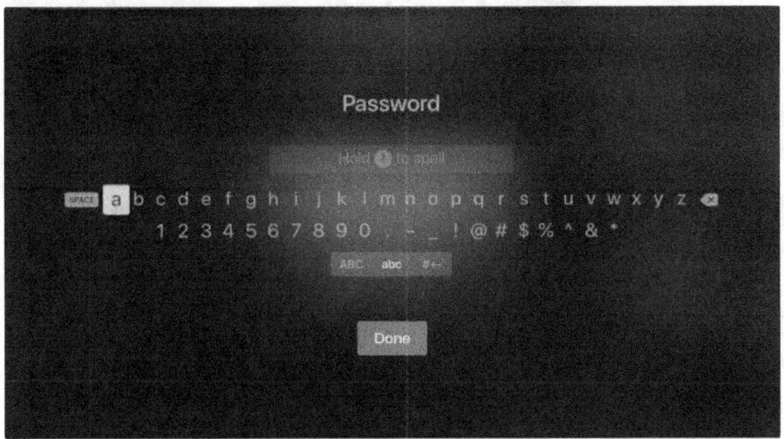

In tvOS 12, coming this fall, the coherence console has been additionally improved to help secret word auto-fulfillment. So at whatever point you discover a login screen, you will see the accompanying warning on your iPhone or iPad

Simply tap on the alert at the most elevated mark of the   to raise the console and you ought to have the option to enter your secret key accreditations on the Apple TV by tapping the AutoComplete suggestion in the Quick Type bar. Also, in case you're utilizing your TV distant or the Control Center interface with your Apple TV, a similar straightforward sign-in measure applies.

The above expects that your Apple TV and your iOS gadget are on a similar iCloud account. But let's say you're a guest at someone's house and you want to access their account on their Apple TV. So what?

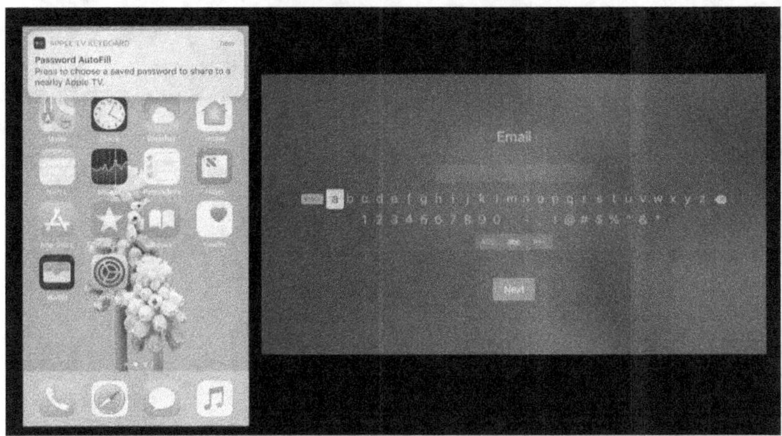

Fortunately, tvOS 12 also fulfills this scenario. Apple TV would now be able to ping the Siri Remote to track down a close by iPhone. When your gadget is distinguished, your iPhone will request that you affirm that you need to utilize the auto-complete element. You will at that point be provoked to enter the validation PIN showed on the close by Apple TV.

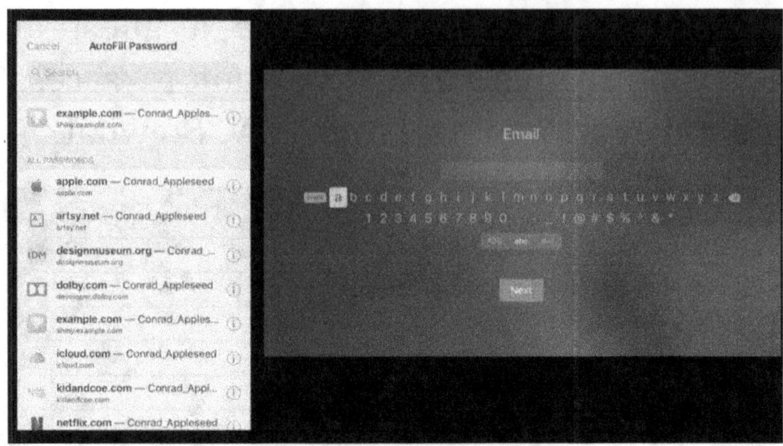

In the wake of confirming your iPhone with Face ID, Touch ID, or a password, you will be given your rundown of passwords and the secret phrase for the pertinent application or administration you need to access will show up at the top, fit to be shipped off Apple. of his companion. Television with a tap.

Programmed password filling on Apple TV requires the establishment of tvOS 12 and iOS 12, which will be delivered in the fall.

Step by step instructions to Pair an Apple Remote with an Apple TV

Instructions to pair an Apple distant with an Apple TV (or even a Mac)

At the point when you set up another Apple TV and turn on the set-top box, the Apple Remote that comes for the situation should normally join when you press one of the buttons. In the event that the Apple Remote quits working, it is most likely dead and should be charged for 30 minutes through a

USB to lightning link associated with a USB outlet.

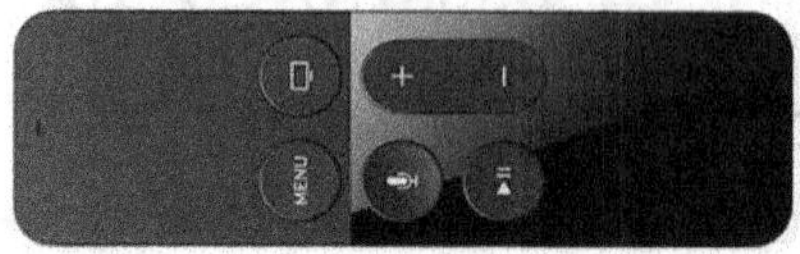

Yet, on the off chance that that doesn't fix the issue, the best activity is to combine the gadget with your Apple TV once more. This article shows you how. The accompanying directions will likewise be useful in the event that you need to combine another substitution Apple Remote in the event that the one that accompanied your Apple TV quits working totally or is damaged beyond repair.

Likewise, toward the finish of this article, we've incorporated a fast tip for matching your Mac with an Apple TV far off, to control things like iTunes, VLC, and Keynote.

Step By Step Instructions To Pair An Apple Remote With Apple TV

Ensure your Apple TV is on.

Point the Apple Remote three creeps from the link box, at that point press and hold the Menu and Volume Up catches on the far off for five seconds.

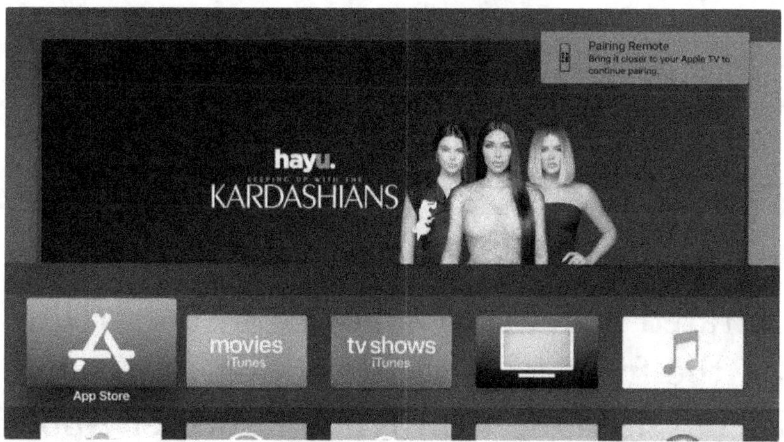

In the event that you see a warning on the TV screen requesting that you bring the Apple Remote nearer, place it on top of the Apple TV.

In the event that you don't see the far off connection notice on the TV screen, unplug your Apple TV from the divider power source, stand by at any rate six seconds, at that point plug it back in.

If important, repeat stages 1 to 3.

Step By Step Instructions To Control Your Mac With An Apple. Remote.

Apple included a little white or silver infrared far off with certain Macs, permitting Mac clients to control things like Keynote introductions and iTunes media from a remote place.

When latest Macs no longer included an infrared beneficiary (showed by a dark line on the front edge of the body), Apple quit including these controllers, however Apple TV proprietors can alternatively utilize their Apple TV far off to control the own Mac, because of an outsider: outsider Bluetooth utility for macOS called SiriMote from Eternal Storms Software.

SiriMote isn't accessible on the App Store, yet you can download it straightforwardly from the Eternal Storms site [Direct link].Subsequent to downloading SiriMote, drag it from the Downloads envelope to the Applications organizer, at that point dispatch the application and adhere to the onscreen directions to match your Apple Remote with your Mac.

Compatibility

tvOS 14 is intended to deal with fourth and fifth era Apple TV models. It isn't viable with more established adaptations of Apple

TV, as these models are not viable with tvOS.

Tvos 14 Timeline

February2021

February 25

Paramount + Channel will cost $ 9.99 per month without ads or $ 4.99 with ads

Last update summary

February 19

Plex Testing TV application joining on iPhone, iPad and Apple TV

February 16

Apple Seeds According to tvOS 14.5 Beta for developers

February 13

CBS All Access Ad Awareness Campaign Reminds Apple TV Owners of Its Rebranding to Paramount + Next Month

February 4th

Deals: 32GB 4K Apple TV Gets Great Discount, Available for $ 149.99 on Amazon ($ 29 Off)

February 1

Apple Seeds first tvOS 14.5 beta for developers

January2021

January 27

Apple heads have examined not 'leaving cash on the table' when choosing Apple TV membership expenses

January 26

Apple releases tvOS 14.4 for fourth and fifth era Apple TV models.

January 13

Apple Seeds tvOS 14.4 second developer beta [Update: public beta available]

December2020

December 22

Rumors continue about a new Apple TV with a stronger gaming focus, an updated remote, and a faster chip next year

December 16

Apple Seeds First beta of tvOS 14.4 for developers

December 15

Nikkei: Apple is working on the new Apple TV for launch next year

December 14

Apple releases tvOS 14.3 for fourth and fifth era Apple TV models

December 8

TvOS Apple Seeds RC Developer Update Version 14.3

December 3

Vodafone Germany supplies Apple TV 4K with each GigaTV contract

2 December

Apple Seeds tvOS 14.3 Third Developer Beta Update [Update: Public Beta Available]

November2020

November 22

Universal Electronics offers an alternative remote control for Apple TV to cable companies

November 18

Apple Seeds tvOS 14.3 Second developer beta update

November 12

Apple Seeds First tvOS 14.3 beta update for developers

November 5th

Apple releases tvOS 14.2 for fourth and fifth era Apple TV models

Part 4

Appletv Tricks Tips: Everything You Didn't Know About Apple TV And Tvos

1 Reorganization Of Applications

If you want to move between apps on your Apple TV home screen, long press on an app with the Siri touchpad, then move it wherever you want.

2 Create Folders

Starting with tvOS 9.2, users can now create app folders. Simply long press one of the applications, as in the past, and drop it onto another application. 3 The Apple Remote application

3 The Apple Remote app

Many don't have a clue about this since it wasn't accessible at dispatch, yet now you can utilize the "Far off" application in the iOS App Store to control your Apple TV. 4

4 Video Settings

When watching a clasp, swipe down from the top to see a rundown of settings and alternatives for your video, including closed captions, AirPlay and Bluetooth speakers, and even a noise reduction mode.

5 custom screen protectors

While the new tvOS aerial screensavers are beautiful, you can also use a photo of yourself. Go to Settings> General> Screen Saver> Screen Saver Type.

6 multitasking

Similar to iOS, the tvOS multitasking app has an app switcher. To perform multiple tasks, double tap the TV button on the controller.

7 sleeping

In the event that you need to take care of your Apple TV mode, simply press the Menu button.

8 Siri

Siri is quite possibly the most remarkable highlights on Apple TV. In addition to asking him for something to watch, you ask him things like sports scores and the weather. You can also take a gander at something and afterward ask Siri "What did she simply say"? Furthermore, it will return to what that individual just said and turn on subtitles for the span of the sentence.

9 Apple Search

Siri can look through various services, including iTunes, Hulu, Netflix, HBO, Fox, CBS, and others, for a particular thing. So in the event that you request that Siri discover you a few comedies, it will show you a few alternatives and afterward what services it is accessible on. You can likewise ask Siri for "clever films on Netflix."

10 Bluetooth Keyboards

Typing on Apple TV can be a problem, anyway tvOS offers full Bluetooth keyboard support. Just go to the Bluetooth options and add the keyboard In the event that you are utilizing an Apple Magic Keyboard, the capacity keys work as well.

11 Voice Dictation

With respect to composing in tvOS, you would now be able to utilize transcription to enter search boxes. Simply press the Siri button while choosing a book box.

12 Bluetooth Headphones

As well as supporting Bluetooth keyboard, tvOS upholds Bluetooth headsets, so you can watch without upsetting others.

13 Viewing The Conference Room

By going to Settings> AirPlay> Conference Room View, you can show a custom message to permit individuals to associate their gadgets to Apple TV and keep the TV from doing whatever else without a password.

14 Watch Live TV

A few applications on the tvOS application store, including CBS All Access, Disney, and FXNOW, support live gushing from their organizations. You can request that Siri go straightforwardly to live transmissions by saying "Watch ____ live."

TvOS 14: update 14.4 is now accessible

Here's what you need to know about tvOS 14, including its top new highlights, delivery date, and how to get beta.

Apple's TvOS isn't your most mainstream working framework, however on the off chance that you own an Apple TV, you're certainly intrigued by what each major new delivery brings.

With tvOS 13 (and later) we got a ton of extraordinary highlights like interface changes, Xbox and PlayStation regulator support, more clients, and Control Center. In correlation, the features available in tvOS 14 are minor, however here and there the seemingly insignificant details have an effect. This is what to expect when tvOS 14 dispatches this fall.

Update 1/26/20: Apple released tvOS 14.4.

The latest: tvOS 14.4 is out

Apple has released an update for tvOS, the working framework for the Apple TV video real time gadget. The tvOS update support archive expresses that "the update incorporates general performance and."improvements."

To download the update, open the Settings application, at that point System, Software refreshes.

How To Get Tvos 14

On Apple TV, dispatch the Settings application and select System, at that point Software Updates. Select Update programming and your Apple TV will check for the latest adaptation. On the off chance that accessible, you can decide to Download and Install or Update Later.

After downloading tvOS 14 and restarting Apple TV, you will be running tvOS 14.

4K Youtube Video

It is astounding that it has taken such a long time. Regardless of what video you watch in the YouTube application on your Apple TV 4K, it will just play at a limit of 1080p without HDR.

With tvOS 14, the "most recent YouTube recordings" will be played in 4K. It's hazy if this is on the grounds that Apple in the end chose to permit the VP9 codec or in the event that it just applies to recordings that YouTube encodes to the new AV1 video design, in any case, it would seem that the most recent YouTube stuff 4K truly do it. .

Picture In Picture Everywhere

With tvOS 13, you can reduce a video to a small square from image to image, but only within the TV app. The tvOS 14 update makes it work at the system level. You can utilize PiP to watch a program while running a preparation application, for instance.

Game Improvements

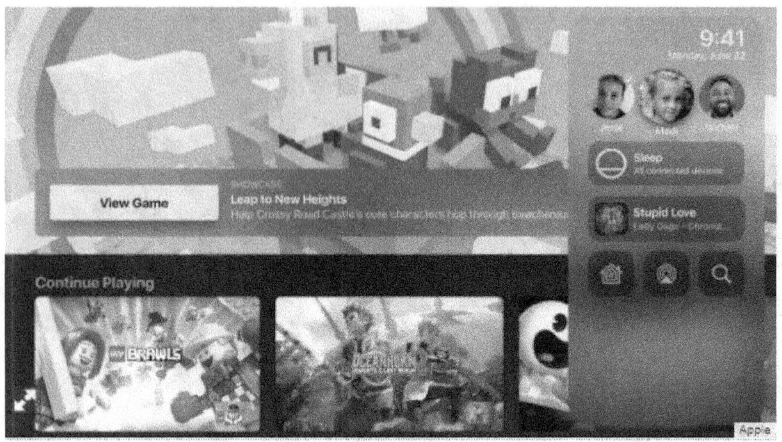

Co-op games on the couch become a real thing with more logins in multiplayer games.

Although tvOS itself has had multi-user support since tvOS 13, it has always been a real pain to switch between users.

It has been fixed in tvOS 14, which allows you to switch users via the Control Center and load the new player's settings and progress, while the first player is saved in the cloud.

Apple is adding support for another pair of Xbox controllers.

Talking about game regulators, Apple has added support for the Xbox Elite form 2 and Xbox Adaptive, notwithstanding the Xbox One and PlayStation 4 regulators it as of now supports.

Airplay In 4K From The Photos App

Your iPhone photographs and recordings will look better than anyone might have expected on your big TV.

At the point when you use AirPlay to send photographs or recordings to your Apple TV 4K from your iPhone (with iOS 14) or iPad (with iPadOS 14), you'll see them in the best quality.

For many people, their 4K TV is the best screen in the house and has a much higher resolution than their iPhone or iPad screen. This will make AirPlay on TV the best way to view photos and videos.

Initial View

Home View allows you to view cameras and shoot scenes.

In the updated Control Center, Apple has a Home View include that can show HomeKit-viable cameras and the sky is the limit from there. It's not the full Home application you have on your telephone, yet it's a method to see cameras and other associated gadgets directly on your TV, just as wake up scenes.

Share Audio

you can connect AirPods to your Apple TV and listen secretly With tvOS 14, you can interface two PCs simultaneously, so you and a companion can tune in without upsetting any other person.

It's a component Apple carried out to iOS and iPadOS a year ago, and it bodes well in the lounge room as well.

Compatible Devices

In case your Apple TV hardware can run tvOS 13, it can run tvOS 14. This implies Apple TV HD (initially called Apple TV fourth era) or Apple TV 4K (fifth era).

In the event that your Apple TV was released before 2015, run old Apple TV programming rather than tvOS and you will not get tvOS 14.

Part 5

Use The Tvos Control Center On Apple TV

Control Center gives you speedy access to settings and controls for exchanging clients, playing music, getting to HomeKit cameras and scenes, taking care of Apple TV, and more. You can quickly switch between clients in the Control Center with the goal that every individual gets their own rundown of recordings, Up Next music and video assortments, Game Center information, and selective suggestions

Open the control center

 Press and hold the Home button On the Siri far off

Switch To Another User

Press and hold the Home button on the Siri remote to open Control Center.

Swipe to feature a client, at that point press the Touch surface to choose it.

At the point when you change to another client, the past client is logged out and the TV and Music applications are refreshed with select records, video or music libraries, and proposals from the new client..

Note: Switching between clients doesn't change the Photos application or different settings related with an iCloud account.

Go to the current song in Music

At the point when a song is played or stopped in the Music application, the Music application on Apple TV shows up in Control Center.

Press and hold the Home button on the Siri remote to open Control Center.

Select the song currently playing.

The Music application opens on the Now Playing screen.

Access the sound controls

Press and hold the Home button on the Siri remote to open Control Center.

Select the Audio Controls button AirPlay sound button, at that point pick earphones or at least one speakers.

Access Scenes And Cameras From Homekit

You can get to HomeKit scenes and any home camera you've set up in the Home application on an iOS, iPadOS, or macOS gadget (with Catalina or later) that is endorsed in with a similar Apple ID.

For instance, you can watch video from viable surveillance cameras and get informed when a viable doorbell camera distinguishes somebody at your entryway. Get an individual warning if HomeKit perceives the individual in your Photos library.

You can likewise play scenes you've made in the Home application on an iOS, iPadOS, or macOS gadget (with Catalina or later). Scenes permit you to control numerous frill in your home. For instance, you can make a Watching TV scene that darkens the lights and sets the temperature.

Press and hold the Home button on the Siri remote to open Control Center.

Select the HomeKit button

Do one of the following:

View live video from a surveillance camera: Select the camera to see a full screen picture or swipe to switch cameras.

Play a favorite scene: Select the scene.

Ask Siri. Say "Show me" trailed by the camera name in the Home application:

"Show me the camera in the yard."

Note: Apple TV can't handle scenes including secure HomeKit viable items like mechanized entryways, entryway or window locks, security frameworks, and carport entryways. You should utilize an iOS, iPadOS, or watchOS gadget to control those gadgets.

Quickly access the search application

Press and hold the Home button on the Siri remote to open Control Center.

Select the search button

Close the Control Center

on your Siri remote ,Press the Menu button

Part 6

APPLE TV SET UP

Set Your Apple ID On Apple TV

Your Apple ID is the account you use for almost everything you do on Apple TV, including buying movies or TV shows, subscribing to Apple TV channels in the TV app in the Apple TV app, also, downloading. Application Store App Store. You can likewise utilize iCloud, which connects you and all your Apple gadgets to share photographs and more.

On the off chance that you don't have an Apple ID yet, you can make one on the Apple ID site. You just need an Apple ID to utilize the Apple TV application, iTunes, iCloud, and Game Center administrations.

Here Are Some Of The Things You Can Do With Your Apple ID On Apple TV:

Apple TV app: Apple TV application: Buy or lease motion pictures, purchase TV scenes and seasons, and buy in to Apple TV + or Apple TV channels inside the application. You can likewise get to your buys made with similar Apple ID on different gadgets.

Music: If you are subscribed in to Apple Music, you can get to a huge number of songs on Apple TV. Furthermore, with an Apple Music membership (or an independently bought iTunes Match membership), you can get to the entirety of your music on the entirety of your gadgets, including music that you imported from CDs or bought from some place other than the iTunes Store.

Apps & Arcade: Buy applications or pursue Apple Arcade straightforwardly on Apple TV and download past App Store buys on Apple TV for nothing, whenever.

Game Center: Play your #1 games with friends who have an Apple TV, an iOS or iPadOS device, or a Mac (OS X 10.8 or later).

Photos: View photographs and recordings from iCloud Photos, My Photos Stream, and Shared Albums

Family Sharing: Share films, TV shows, applications, and bought Apple TV memberships with up to six relatives.

A Home Screen - Keep introduced applications and Home screen something similar on each Apple TV you own.

Sign in with Apple: Sign in to applications with your current Apple ID, without rounding out structures or making new passwords. Apple doesn't follow your action and you are in charge of your information.

AirPods Compatibility - Listen with AirPods, no arrangement required. AirPods linked to your Apple ID automatically connect to your Apple TV.

Note: not all featuress are accessible in all nations/areas.

Set password prerequisites for buys

You can set whether Apple TV requires your Apple ID secret phrase to finish a buy from the iTunes Store or App Store.

Open Settings Settings on Apple TV.

Go to Users and Accounts> [account name]> Purchases, at that point select Always, following 15 minutes, or Never.

Allow Free Downloads

You can set whether Apple TV requires your Apple ID secret key to permit free downloads from the iTunes Store or App Store.

Open Settings Settings on Apple TV.

Go to Users and Accounts> [account name]> Free Downloads, at that point select Yes or No.

Presently Playing: Control music playback on Apple TV

Regardless of where you are in the Apple Music App Music application, you can choose a tune and afterward press the touch cushion to begin playing it.

When a song begins to play, it is displayed in Now Playing.

Song playback continues even if you exit Music, but stops if you start playing video or audio in another application.

Go To Now Playing

Open the Music Music app on Apple TV, then scroll down to Now Playing.

Tracks adjacent to the one currently playing are displayed in the queue. A timeline is also displayed showing the elapsed and remaining time. At the point when the timetable is dynamic, you can likewise press the Touch surface to play or interruption the melody.

Once a song has started playing, the display changes to show only the song currently playing. Press the Menu button button to return to the queue.

Control The Music While It Is Playing

With the Now Playing screen open to line on Apple TV, swipe left or right to the melody you need to play, at that point press the Siri Remote's touch cushion to begin playing it.

The line changes to show just the tune at present playing.

Do one of the following on your Siri remote:

Pause or Play: Press the Play/Pause button Play/Pause catch or press the touch cushion

Return to the start or jump to the following song: Press the left or right half of the Touch surface to restart the current tune or jump to the beginning of the next song.

Rewind or fast forward: Press and hold left or right to rewind or fast forward during playback. Release to resume playback.

Move to a specific point in the song: Press the Touch surface to pause the song and show the playhead, then slide your finger left or right to go forward or backward in the timeline. Press the Touch surface again to continue playback.

Back to queue: Press the Menu button.

Use the "Now ringing" queue

The line shows all the songs in a collection or tunes and recordings in a playlist. Lined things are shown in succession with the as of now playing melody in the middle.

Do one of the following:

Explore the line: Swipe left or right on the Touch surface.

Start playing an alternate tune: Highlight a lined tune and afterward press the Touch surface.

Add An Item Immediately After The Currently Playing Item: Highlight a melody in the line, look down and select the more catch at the lower part of the screen, at that point select Play Next.

Stream content playing on Bluetooth or AirPlay viable gadgets

For more data on playing sound on different speakers or earphones

Do one of the following:

With the Now Playing screen open on Apple TV, swipe up to the controls at the highest point of the screen; Scroll left and select the Audio Controls button AirPlay Audio Controls button.

Press and hold the Home catch on the Siri far off to open Control Center, at that point select the Audio Controls button AirPlay Audio Controls button.

Select at least one playback objections.

To sort out some way to connect a Bluetooth device, see Connect Bluetooth Devices to Apple TV. For more data on AirPlay streaming.

Quickly Add A Song To Your Library

With the Now Playing screen open on Apple TV, swipe up and select the Add button Add button.

Look At The Lyrics Of A Song

You can view the lyrics of a song (enabled by default) by subscribing to Apple Music.

With the Now Playing screen open on Apple TV, ensure the Lyrics button Lyrics button in the upper right corner is chosen.

Select a song in the queue.

While the song is playing, the lyrics are displayed and move to the beat of the song.

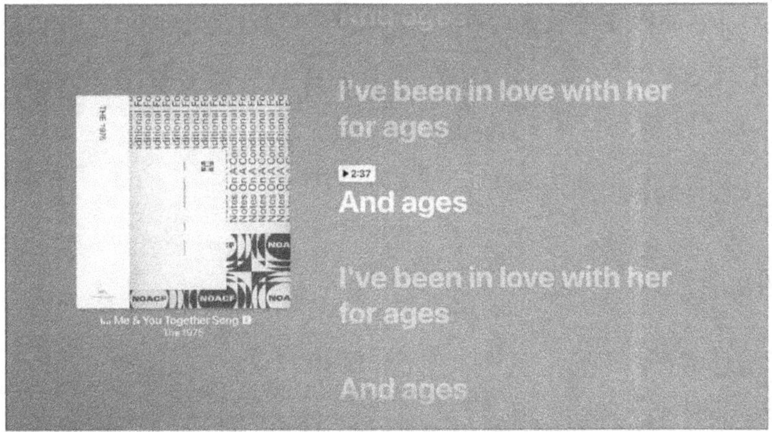

To find the letter you want to hear, slide your finger up and down the letter.

As you slide your finger all over, a catch seems to demonstrate the particular time each letter can be heard in the melody's timeline. Press the Siri remote's touch pad to start playback from there.

To turn off lyrics, , look up and select the Lyrics button Lyrics button at the highest point of the screen.

Note: The Lyrics button will not appear if lyrics are not available for the song currently playing.

With the line open on Apple TV, look down and select the More catch at the lower part of the screen. Dependent upon the tune, you may have the choice to:

Go to the album

Go to the artist

Add the tune or eliminate the tune from your library

Add the song to a playlist

Play the next song

Start a custom radio broadcast from the song

See the full lyrics of the songs

Mark the song as one you love or don't love

Suggest fewer songs like this

Repeat the song currently playing or the entire Now Playing queue

With the Now Playing screen open on Apple TV, swipe up and select the Repeat button Repeat button at the highest point of the screen.

To repeat the entire Now Playing line, select the Repeat button once again.

To turn off repeat, select the Repeat button Repeat button one more time.

Turn Auto Play On Or Off

Autoplay (turned on by default) keeps expanding the queue with similar songs until you stop playing the music. You will be able to see the next ten songs in the queue.

With the Now Playing screen open on Apple TV, swipe up and select the Repeat button at the highest point of the screen.

To reactivate AutoPlay, select the AutoPlay button again. AutoPlay button.

Mix Songs

With the Now Playing screen open on Apple TV, swipe up and select the Repeat button at the top of the screen.

To turn off shuffle play, select the Shuffle once more.

Search Apple TV

Apple TV consolidates a pursuit application Search App that can help you with discovering films, TV shows, cast and group information, applications in the App Store App Store, and even music in case you're an Apple Music subscriber.

Search utilizing the on-screen keyboard

Feature the Search App Search application on the Apple TV home screen, at that point press the Siri Remote Touch Pad.

Enter the query using the on-screen keyboard.

The search application returns the content associated with the search terms.

Tip: You can utilize the keyboard on a close by iPhone or iPad to enter text directly on Apple TV as opposed to utilizing the Siri Remote.

Dictate Instead Of Write

You can also enter search terms in the search application using voice dictation.

Press and hold the Siri button Siri button on your Siri far off, at that point talk.

Talk To Your Apple TV

Siri makes cooperating with Apple TV simple, fun, and educational. You can look for films, TV shows, music or applications; discover entertainers or chiefs you like; control playback; open your applications;

and even get some information about sports, climate and activities, paying little heed to what occurs on the screen.

Siri doesn't react to you on Apple TV like on iPhone and different gadgets, yet it will satisfy your solicitation and show the outcomes on the screen.

See A Rundown Of Things You Can Ask Siri

Press the Siri button Siri button on the Siri remote.

Find and control Apple TV with your voice

Press and hold the Siri button Siri button on the Siri remote and start talking.

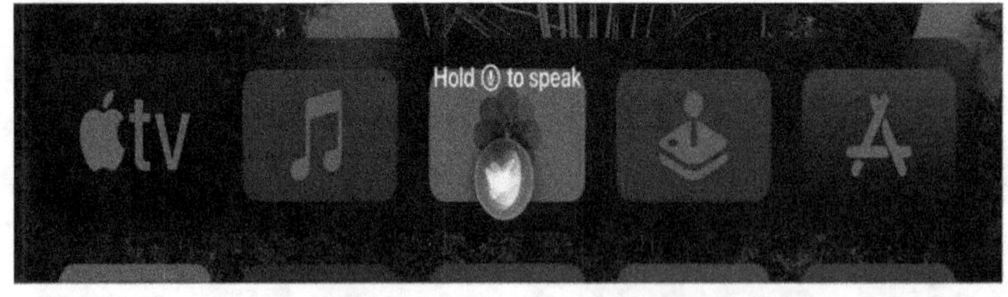

Siri comprehends a wide scope of orders and naturally applies them to the current setting at whatever point conceivable.

Dictate Instead Of Writing

At whatever point you see a book input field, you can utilize your voice as opposed to using the one on-screen keyboard.

Press and hold the Siri button Siri button on the Siri far off and talk the content you need to embed. You can likewise talk singular characters, for instance when entering client names and passwords.

Select A Search Result

Select an item from the results list to see more details.

Data about the item is shown, including all applications where the substance is accessible.

Sleep or wake up Apple TV

Apple TV is ready to watch at any time and automatically goes into standby mode after a preset period of inactivity.

Wake Apple TV From Sleep

Press the Menu button, Start Home button, Siri Siri button, or Play / Pause Play / Pause button on the Siri remote.

Apple TV is prepared to watch whenever and consequently goes into reserve mode after a preset time of inertia.

Set The Delay Before The Suspension Begins

To set how long Apple TV waits before going to sleep, open Settings Settings on Apple TV.

Go to General> Sleep Later and choose an option.

Force Immediate Suspension Of Apple TV

Do one of the following:

On Siri Remote, press and hold the Home button Home button to open Control Center, then select Sleep.

Open Settings Settings on Apple TV, then select Sleep Now.

When you're done, you can also turn off your TV or AV receiver. Apple TV falls asleep after a time of idleness